FOI

I remember the first time that ~~I met Codie-lyn~~ in her home. I traveled from Oklahoma to Ohio to teach anoint her family with Young Living Essential oils and the Raindrop Technique. She gave me her full attention and asked very intelligent questions about the oils. I noticed she wasn't taking any notes. I was concerned that she wouldn't remember all the things I was saying, but she told me that she remembers everything that she reads and would remember everything that I was teaching her.

Over the next few years, Codie-lyn proved to me that her "bulldog" tenacity and "momma bear" belief system was definitely the combination that has made her a great example of a Biblical Proverbs 31 woman. I am proud that this accomplished Diamond is part of my Young Living Organization. Not only did she use Young Living Oils to help her sons overcome all medical adversities that they faced, but she continued searching for the "missing" link that would excel their DNA transformation from the "drug" damage to their young bodies.

She has been using CBD oil with her children for over a year now. I trust her research into the vast ocean of CBD experts and over-night companies. and stores that have literally posted billboards on every corner in seemingly every city or town in America. The over-whelming "gold rush" into the CBD oil phenomena has been daunting to say the least. The questions and responses from several Young Living Distributors have also been diverse and confusing.

I thank God that I know and trust Codie-Lyn. She is helping me make my way through my own health challenges with CBD. I also want to thank Young Living that provided the essential oils that initially brought Codie-lyn and I together. Without their support with our own pure therapeutic-grade essential oils and recommended source of pure CBD, I would have never entered into this new frontier. Once again, thank you so much Codie-lyn.

Karen Hopkins
Young Living Royal Crown Diamond

INTRODUCTION

My name is Codie-lyn Kahler. I'm a wife, business owner, and a Mom-on-a-mission (MOM). I have been blessed with amazing children, but our journey so far has included many ups and downs as we have overcome significant health issues. Through it all, I have had to learn to be an advocate for my children.

In 2015, we embarked on a gluten-free lifestyle after my son began having severe reactions to gluten. These reactions triggered a neurological response, which caused tremors and stomach pains along with diarrhea and vomiting. In time, we learned that we are carriers of the Celiac gene mutations.

Celiac Disease is thought to be an autoimmune condition that triggers an immune response to gluten, causes damage to the small intestine, and can even cause damage to the brain. A gluten-free diet helped to stop symptoms, and our daily use of therapeutic grade essential oils helped to support healthy digestion. However, if accidental gluten ingestion occurred, the tremors would return, along with the digestive upset and pain.

In February 2018, we decided to try CBD. Within a few months we noticed that the tremors disappeared and did not return, even with accidental gluten ingestion. We also noticed that the digestive reaction was significantly reduced.

Two studies have been done with CBD, cannabinoid receptors, and Celiac disease. They found that CBD can have a therapeutic role in treating Celiac Disease, which helps to explain the positive results we experienced firsthand [1][2]. After seeing such improvement in our family, I decided to research CBD and the vast benefits of this powerful plant compound even further.

CBD OIL

We live in an age of great scientific exploration. Technology has allowed us to dig even deeper into the human body and gain more understanding of the complexities of how it functions. Scientists are continually learning and improving upon their theories in an attempt to make humans as healthy as possible. In 1940, a compound named Cannabidiol (CBD) was discovered. This led to the 1992 discovery of another system of the body, the endocannabinoid system [3].

Since the discovery of the endocannabinoid system, researchers have a better understanding of how it works and the plant-

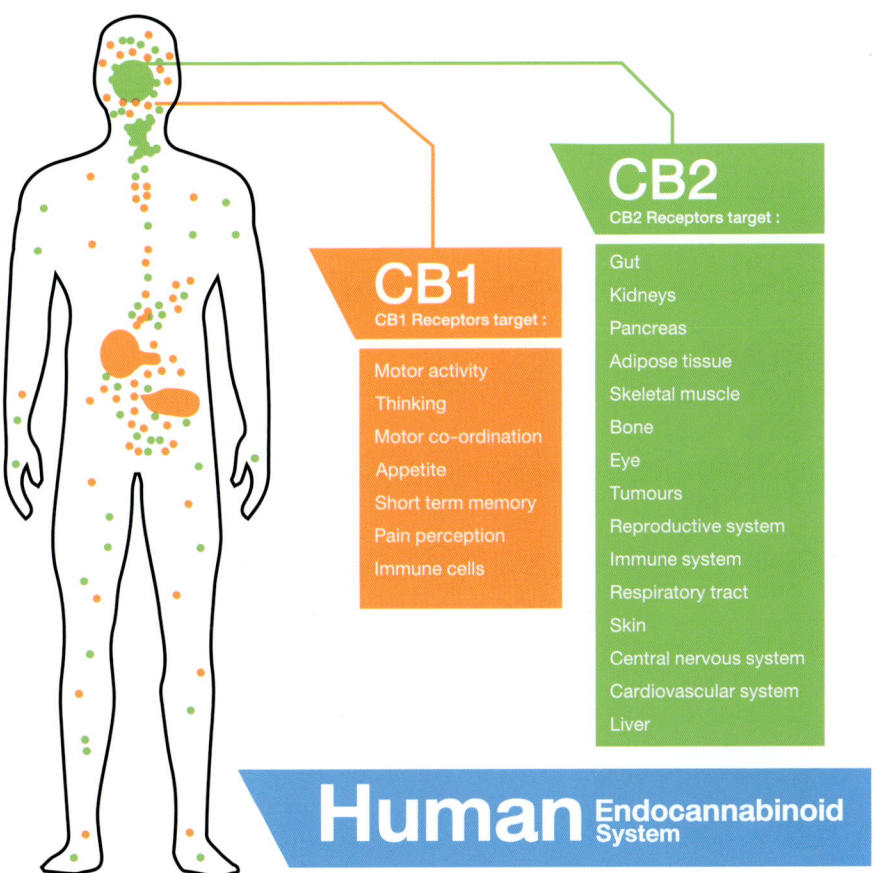

based cannabinoids that have an impact on overall health and wellness. The endocannabinoid system helps to regulate the nervous, endocrine, and immune systems along with functions of the brain. It helps with things like digestion, pain modulation, reproduction, stress, appetite control, mood, temperature control, and even memory [3]. This complex system is still undergoing further research and has become the new frontier of plant-based medicines. In order to understand the benefits of the endocannabinoid system, it is important to know how this system works.

When your brain wants to send a message to the body, it releases a chemical called a neurotransmitter. This neurotransmitter is released from a neuron (presynaptic). Once this message is released, it must reach across a gap, called a synapse, to the next neuron (postsynaptic). When the message is received, it excites that neuron, which then allows the message to continue moving from neuron to neuron until it reaches its final destination.

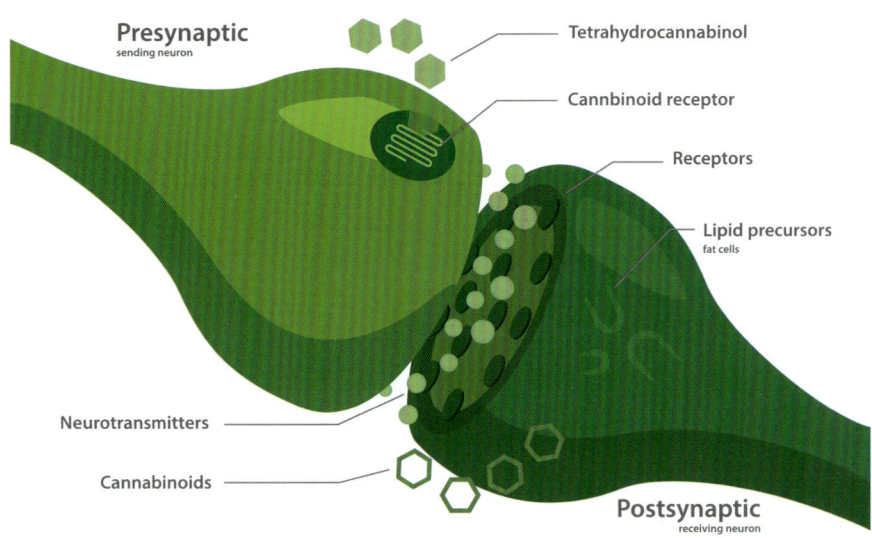

The Endocannabinoid System Explained

Think of this kind of like the old-fashioned game of telephone where a group of children sit in a circle and one person whispers a message into the ear of the next and the next until it reaches the end of the circle. The goal is for the message to stay the same from person to person, but it often gets messed up and by the end, it is a very different message than how it started.

When the neurotransmitter excites the neuron, it interacts with cannabinoid receptors acting as locks that can only be unlocked with a specific set of keys. Two of the most well-known receptors in this system are CB1 and CB2. The CB1 receptors are found abundantly in the central nervous system. The CB2 receptors are typically found in immune cells, the gastrointestinal tract, and the peripheral nervous system.

When the postsynaptic neuron receives a message, it activates these cannabinoid receptors and sends a message back the opposite direction. It often limits the response of the first neuron so that less of the neurotransmitter is released [4]. In the game of telephone, it would be like whispering back to the person that gave you the message and correcting what was passed on. This action effects how the messages are sent, received and processed, and it also helps to regulate the processes of the body creating homeostasis, or balance.

The receptors that are stimulated in this process interact with two main cannabinoids: Tetrahydrocannabinol (THC) and Cannabidiol (CBD). THC comes from the marijuana plant, *Cannabis Indica* or *Cannabis Sativa*, and gives a euphoric feeling often referred to as "getting high" [5]. THC is the cannabinoid that interacts with CB1 receptors, mainly in the brain and central nervous system. On the other hand, CBD comes from the hemp plant, Cannabis Sativa and does not interact with the body in the same way; therefore, there is no euphoric feeling or altered mental state [5]. CBD works with CB2 receptors and has more of an impact on the immune, gastrointestinal, and peripheral nervous systems [6].

Historically, hemp has been grown and used for centuries for everything from cloth and paper to lamp fuels, ropes, industrial fiber, food, and even body care products. During Colonial times, everyone was mandated to grow hemp [7]. Presidents Washington and Jefferson both grew hemp, and it was widely grown in the US up through the mid-twentieth century [8]. It wasn't until 1970 that congress passed the Controlled Substances Act and classified hemp as a Schedule 1 drug

with strict regulations for its use and cultivation [9]. In 2018, congress passed the Agricultural Improvement Act that removed hemp and some cannabis products from the Controlled Substances Act, thereby making them more readily available. Currently, hemp is legal in all 50 states [10]. Marijuana remains a Schedule 1 drug in the United States and is only legally available in some areas of the country.

CBD was first researched by Dr. Walter S. Loewe in 1946 [11]. He tested the compound on animals and found it did not create an altered mental state. Research continued into the 1960's when British Pharmacopoeia released a CBD oil for therapeutic use [12]. Then in 1980, Dr. Raphael Mechoulam found that CBD could be a key factor in treating epilepsy [13].

CBD PROCESSING

Today, CBD is obtained from the hemp material through a process called Supercritical CO_2 Extraction [14]. This process puts CO_2 under high pressure while staying at a low temperature. The high pressure turns the CO_2 into a liquid that passes through the plant material and extracts pure CBD oil. This process is 90% efficient.

To create a CBD Isolate, this process goes even further. Once the oil is released in the extraction process, chemists use chromatography to remove other cannabinoids and plant material. Then the final step is heating and activating the CBD in a process called decarboxylation. The result is a CBD Isolate that is over 99% pure with 0.00% THC [15]. As with most industries, there are always imitators trying to profit from the therapeutic benefits while providing a sub-optimal product. To ensure the safety, purity, and therapeutic value of CBD oil, it is important that it be tested and free from heavy metals, pesticides, and solvents.

CBD ISOLATE

A CBD Isolate is a white crystalline powder that contains 99% CBD and 0.00% THC [16]. This means the product can be used in many ways such as nutritionally, topically, and internally. Because it is an isolate, there are no trace amounts of THC, which means it will not interact with drug testing.

A full-spectrum CBD oil is one that still contains plant material, chlorophyll, waxes, oils, terpenes, and various other cannabinoids. The down side to a full-spectrum oil is the risk of getting drug test results that are positive for trace amounts of THC. The benefit of a full-spectrum oil is the terpenes that create a synergistic effect in the body when combined with CBD. We can get the benefit of these terpenes by simply adding terpene-rich therapeutic-grade essential oils to a CBD Isolate.

TERPENES

What are terpenes?
Terpenes are aromatic hydrocarbons that are found in the essential oils of a plant. They effect the CB1 and CB2 receptors and neurotransmitters in the brain and mimic antidepressants. A 2011 article published in the British Journal of Pharmacology found that the complex interaction between terpenes and cannabinoids like CBD and THC "could produce synergy with respect to treatment of pain, inflammation, depression, anxiety, addiction, epilepsy, cancer, fungal and bacterial infections" [17]. Some of the most common terpenes are Myrcene, Limonene, Caryophyllene, Terpineol, and Linalool.

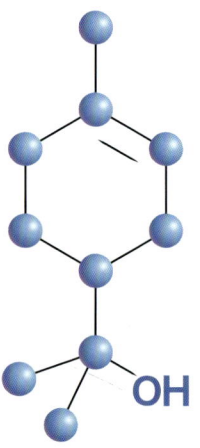

Terpineol

Here is a chart that shows the effect of each terpene:

MYRCENE	CARYOPHYLLENE	TERPINEOL	LINALOOL
Relaxing	Anti-fungal	Anti-bacterial	Sedative effect
Anti-mutagenic	Anti-bacterial	Antioxidant	Anti-anxiety
Anti-inflammatory	Citrusy Scent	Sweet smelling scent	Analgesic
Earthy scent	Helps with depression and Anxiety	Bronchodilator	Anti-inflammatory
Muscle Relaxant	Gastroprotective	Aids Memory	Floral Scent

*Information from *The Chemistry of Essential Oils Made Simple: God's Love Manifest in Molecules*, David Stewart 2005

Many therapeutic grade essential oils are high in terpenes. A therapeutic grade oil is one that is tested for purity and potency, and one that has the highest standards of sourcing. Young Living Essential Oils has been around for over 25 years and is the industry leader in essential oils providing therapeutic grade oils and oil-infused products. These terpene-rich oils have been used for centuries to increase overall health and wellness.

Research done by University of Siena in Italy found that the "potential effects of terpenes on cancer, either alone or in combination with cannabinoids, have not yet been addressed in laboratory studies. Indeed, the synergistic effect between cannabinoids and terpenes is often claimed to be the major difference between 'holistic' herbal preparations of cannabis, and products based on single cannabinoids" [18].

When these terpene-rich essential oils are added to a CBD Isolate, they create a synergistic effect in the body. Essential oils can contain several different types of terpenes including monoterpenes, diterpenes, and sesquiterpenes. Below is a chart of essential oils that are terpene-rich. These would be excellent oils to use to enhance your CBD isolate product. They can be added to a CBD oil, used topically, or diffused while using a CBD product.

MONOTERPENES	DITERPENES	SESQUITERPENES
Grapefruit	Pine	Cedarwood
Orange	Cistus	Patchouli
Balsam Fir	Cypress	Sandalwood
Angelica	Clary Sage	Ginger
Lemon		Blue Cypress
Frankincense		Myrrh
Celery Seed		Vetiver
Elemi		German Chamomile
Nutmeg		Black Pepper
Lime		Spikenard
Bergamot		Copaiba
Juniper		
Cistus		
Tea Tree		
Pine		
Black Pepper		
Blue Tansy		
Dill		
Spruce		

*Reference material from *Essential Oils Desk Reference 5th Edition*, Lifescience Publishing 2011

CBD APPLICATIONS

CBD oil is a versatile product that can be used in many different ways. Before dosing yourself with CBD oil, it is important to decide which application method will be best for your specific needs. Some application methods are more effective than others at targeting specific issues. Let's look at the three main application methods:

Topical
This method can be used for things like skin issues and for targeting muscle and joint pain. The CBD absorbs into the skin but very little, if any, enters the bloodstream. It can be used as a spot treatment or partnered with oral application for an even bigger impact. CBD Isolate can be mixed into lotions, creams, and salves for topical usage.

Oral (Swallowed)
This method directs the CBD to the bloodstream by way of the digestive system. It is best to take CBD on a full stomach as it increases absorption fivefold. It is also important to pair it with healthy fats such as MCT oil. Doing this increases the absorption

threefold. This is why CBD Isolate is mixed with a carrier oil. CBD Oil taken in this manner is good for digestive and immune system issues. Peak bloodstream levels can be seen from 30 minutes to five hours after use.

Oral (Sublingual)
This method also directs the CBD directly into the bloodstream. CBD oil is placed under the tongue and held for about 1.5 minutes. This allows the CBD to absorb into the capillary-rich environment, and individuals can see a more immediate impact. CBD oil can also be swished in the mouth for that timeframe, increasing the surface contact between the oil and capillaries. Peak bloodstream levels can be seen from 30 minutes to five hours after use.

For most people starting a CBD regimen, it is best to start with a smaller amount and work your way up, increasing incrementally until you find your optimal dose. It is more important to take the product consistently than it is to take larger doses infrequently. Most individuals can start out with about 5 drops (1/4 dropper) twice a day for a week. After this, the amount can be increased to 10 drops (1/2 dropper) twice a day for one to two weeks. The amount can be adjusted up or down based on your specific needs.

CBD can be used topically and orally at the same time. The amount of CBD a person needs will vary by individual. For individuals with more moderate health concerns, you would start with the lower dose and ramp up more quickly. For significant health concerns, you would start with a higher dose and gradually increase until you find your desired amount.

In clinical trials, participants were given 1,200mg doses daily for several months without any significant side effects [19]. Typically, the only side effects seen were change in appetite, diarrhea, and fatigue [20]. It is important to note that CBD oil inhibits cytochromep450 in the same manner as grapefruit. This means that it will interact with some medications. If your medication carries a warning against using with grapefruit, then you will also want to avoid CBD oil.

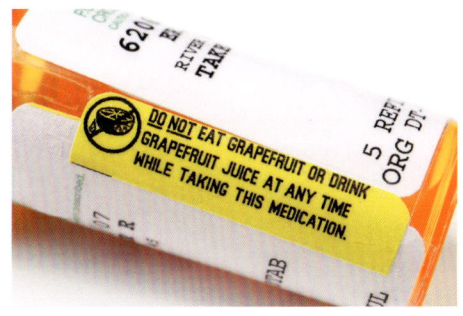

THERAPEUTIC GRADE ESSENTIAL OILS

To increase the impact of CBD Isolate, therapeutic grade essential oils can be added or used in combination. As with CBD dosing, you want to consider the application method that would make the most sense for your health concern. A good reference book for essential oils can help you to see which terpene-rich oils might be best for your specific need and how to apply them. Therapeutic grade essential oils can be used in a similar way as CBD. There are three primary application methods: Topically, Internally, and Aromatically.

TOPICAL
Therapeutic grade essential oils can be applied directly to the skin. For skin conditions, you would apply to the area of concern. They can also be used topically for muscle and joint pain and for relaxation. Some oils may create a warm sensation on the skin and are best diluted with a carrier oil when applied. It is best to dilute oils when applying on young children as well. If you use essential oils with your CBD oil topically, the MCT oil in the CBD will act as a carrier oil. Using the carrier oil will increase absorption of the essential oil.

Black Pepper oil would be a great oil to apply topically with CBD oil as it contains analgesic compounds. This could be used for chronic pain, arthritis, joint pain, and sore muscles. Oils such as Copaiba, Frankincense, Balsam Fir and Cypress can be applied topically with CBD oil to decrease inflammation. Because essential oils are so diverse, there are endless combinations of essential oils with CBD that you can create to find what works best for you.

INTERNAL
Therapeutic grade essential oils can also be used internally. For this method, I only recommend the use of Young Living Essential Oils (YLEOs). It is imperative that essential oils be tested to ensure purity. Young Living has a strict Seed to Seal® commitment that guarantees the oils will be of the highest quality, tested for purity, and grown and harvested in a sustainable manner.

YLEOs can be taken in capsules, added to food or drinks, placed under the tongue, or taken in oil-infused supplements. When taken with CBD, the essential oils will increase the effect of the CBD in the bloodstream and help to target specific needs. Nutmeg oil can be used with CBD to reduce inflammation and pain, improve digestion, reduce fatigue, and support adrenals. Orange oil can be taken with CBD to boost immunity, reduce depression, support heart health, and aid in relaxation.

AROMATIC
This method of application involves breathing in the essential oils. This can be done a couple of ways. You can place a drop of essential oil into your hand and cup over your nose before breathing deeply. You can also use a diffuser to get oil molecules into the air.

When essential oils are breathed into the nose, they trigger the olfactory nerve which sends a signal to the limbic system located in the back of the brain. Inhaled oils can show up instantaneously in the hypothalamus. The oil molecules are so small that they are able to cross the blood-brain barrier and have a positive impact on the brain. Oils high in sesquiterpenes are also excellent for brain health. These oils can be inhaled while using CBD sublingually for optimal benefit.

Cedarwood can be used for calming and to stimulate melatonin production for sleep. Sandalwood stimulates the immune system as well as the pineal gland and limbic systems. In can be used with CBD to enhance deep sleep and improve immune function.

These are just a few of the ways that essential oils can be partnered with CBD for a lasting impact. Remember that CBD can be used topically and orally at the same time, just like essential oils can be used topically, orally, and aromatically at the same time.

WHY USE CBD OIL?

You may be wondering the benefits of this powerful product. Science has a lot to say on this topic. One thing to consider when researching CBD oil is whether the study looks at CB1 or CB2 receptors. If a study is only looking at CB1 receptors, then the product they are testing is medical marijuana and the cannabinoid THC. Some studies test CBD and THC together to see the synergistic benefits of the combination on the body, while others just test CBD.

It is important to remember that CB1 receptors primarily effect the central nervous system and the brain. CB2 receptors mainly effect the digestive and immune systems, as well as the peripheral nervous system. Here are some things we know about CBD and the impact on CB2 receptors, based on scientific studies.

When CBD targets the endocannabinoid receptors, it regulates the neurotransmitters and the response of neurons [21]. This means it is very good at reducing chronic pain and

inflammation [22]. It has been found to reduce sciatic pain and inflammation as well. In one study, they found CBD decreased the pain at the site of surgical incisions [23].

In men, CBD has been shown to accelerate spermatogenesis and regulate the creation of sperm [24]. In women, it has been shown to assist with pain and inflammation throughout the menstrual cycle and can be used vaginally to treat endometriosis and menstrual cramping [25]. Rectal use is another method to target specific needs in both men and women. CBD oil has also been used as a stimulant and lubricant for sexual pleasure.

Because of the impact in the digestive system, CBD can increase serotonin, and thereby increase mood and reduce anxiety. Subjects of one study took 600mg of CBD daily to reduce social anxiety [26]. This is a common symptom many women experience after having children, due to fluctuating hormones. Endocannabinoids can be found in breast milk, and CBD is generally thought to be safe for pregnant women and nursing mothers.

Through research, chemists have found that CBD mitigated the euphoric side effects of THC when used in combination [27]. When CBD is used alone, it can have antipsychotic effects and has even been shown to decrease morphine dependence and heroine-seeking behavior [28][29][30].

CBD has also been tested with some of the most chronic health conditions that we see in the US. It was shown to reduce risk of diabetes by 56% in non-obese individuals and decrease blood pressure [31][32]. Male subjects were given 600mg a day and found a decrease in resting blood pressure [33][34]. The anti-anxiety effect of CBD can also work to decrease high blood pressure caused from stress.

CBD is not only good for chronic conditions but also for daily issues that arise. It can reduce stress, improve quality of sleep, and help balance digestion [35]. It can also be used topically to reduce acne in both men and women [36][37].

The most exciting research being done is looking at the impact of CBD and THC on neurodegenerative diseases and cancer. So far, they have found that CBD significantly decreases seizure activity in patients with Dravet Syndrome [38]. It also increases the quality of life in Parkinson's patients [39][40]. Studies have shown that CBD decreases the inflammation and helps prevent neurodegeneration in Alzheimer's patients. In mice predisposed to Alzheimer's, they found CBD was able to prevent cognitive decline [41][42].

For cancer patients, CBD was able to reduce symptoms associated with treatment such as nausea, vomiting, and pain [43]. In one test tube study, they found that concentrated CBD induces cell death in human breast cancer cells [44]. Another study showed that CBD inhibited the spread of aggressive breast cancer cells in mice [45][46]. More research needs to be done in these areas, but the results so far are very promising.

The endocannabinoid system is the new frontier for scientific research. Studies are continually being done to discover more of the benefits of both CB1 and CB2 receptors, along with the health benefits and overall impact of cannabinoids from plants. Here are a few of the more recent studies that have been done.

- A study published in the 2006 edition of *Endocrine Reviews* says, "The ability of the endocannabinoid system to control appetite, food intake, and energy balance has recently received great attention, particularly in the light of the different modes of action underlying these functions" [47].

- A 2008 study titled *Targeting CB2 receptors and the endocannabinoid system for the treatment of pain* says "CB2 receptors were originally described as being restricted to cells of immune origin, but there is evidence for their expression in human primary sensory neurons, and increased levels of CB2 receptors reported in human peripheral nerves have been seen after injury, particularly in painful neuromas. CB2 receptor agonists produce antinociceptive effects in models of inflammatory and nociceptive pain, and in some cases these effects involve activation of the opioid system" [48].

- A 2011 study published in the *Journal of Psychopharmacology* states, "Thus, both for anxiety and fear memory processing, endocannabinoid signaling may ensure an appropriate reaction to stressful events. Therefore, the ECS can be considered as a regulatory buffer system for emotional responses" [49].

- A 2003 study titled *Activation of CB2 cannabinoid receptors by AM1241 inhibits experimental neuropathic pain: Pain inhibition by receptors not present in the CNS* says, "These findings demonstrate a mechanism leading to the inhibition of pain, one that targets receptors localized exclusively outside the CNS. Further, they suggest the potential use of CB2 receptor-selective agonists for treatment of human neuropathic pain, a condition currently without consistently effective therapies" [50].

- A 2013 review by Silvestri and Di Marzo explains, "In this review, we discuss how, in addition to promoting energy intake, endocannabinoids control lipid and glucose metabolism in several peripheral organs, particularly the liver and adipose tissue. Direct actions in skeletal muscle and pancreas are also emerging. This knowledge may help in the design of future therapies for the metabolic syndrome" [51].

- In the 2008 study *Modulation of the Endocannabinoid System in Cardiovascular Disease* it states, "However, the ECS may also be activated as a compensatory mechanism in various forms of hypertension where it counteracts not only the increase in arterial pressure, but also the inappropriately increased cardiac contractility through activation of CB1 receptors. In addition, the activation of CB2 receptors in endothelial and inflammatory cells by endogenous or exogenous ligands was found to limit the endothelial inflammatory response, chemotaxis, and adhesion of inflammatory cells to the activated endothelium with the consequent release of various proinflammatory mediators, which are key processes in the initiation and progression of atherosclerosis and reperfusion injury as well as smooth muscle proliferation. Therefore, depending on the underlying pathology, selective activation of CB1 or CB2 receptors or inhibition of CB1 receptors may offer therapeutic benefits. It also states, "The majority of findings suggest that the ECS plays a limited, if any, role in cardiovascular regulation under normal conditions. However, it may emerge as an important player under various pathological conditions, ranging from hypertension and the metabolic syndrome to various forms of circulatory shock, atherosclerosis and restenosis, myocardial infarction, and heart failure, as discussed in more detail subsequently" [52].

CONCLUSION

As you can see, there are a variety of ways to use both CBD oil and therapeutic grade essential oils in a safe and healthy manner. When used in combination, they create something even more powerful and life-changing. The full array of benefits of CBD and therapeutic grade essential oils are still being researched and discovered. Scientists are now able to see why these plant-based products have been used for hundreds of years with very few, if any, side effects. I encourage you to make these products part of your daily wellness routine and experience the benefits first hand.

YOUNG LIVING ESSENTIAL OILS

Getting started has never been easier! Young Living Essential oils offers a membership discount when you purchase one of their lifestyle starter kits. The Premium Starter Kit contains 12 oils, 11 of which contain terpenes that enhance the effect of the CBD Isolate.

ESSENTIAL OIL	LIMONENE	BETA-CARYOPHYLLENE	TERPINEOL	MYRCENE	LINALOOL
Peppermint Vitality™					
Lemon Vitality™	X		X		
Citrus Fresh™ Vitality™	X		X	X	
Thieves® Vitality™	X	X	X		X
Digize™ Vitality™	X		X	X	
Raven™	X		X		
Valor®	X	X	X	X	
PanAway®		X			
Lavender			X		X
Frankincense	X	X	X	X	
Stress Away™	X	X			
Peace & Calming™	X	X		X	X

*Chart compiled by Codie-lyn Kahler with reference material from *Essential Oils Desk Reference 5th Edition*, Lifescience Publishing 2011

The Starter Kit also contains a diffuser, samples of Thieves® household cleaner, samples of the antioxidant juice drink Ningxia Red®, and comes with a 24% discount on purchases. To get started:

1. Make a list of the products you would like to purchase. A full catalog of the products can be found at www.youngliving.com

2. Have a credit/debit card or check handy.

3. Call Young Living at 1-800-371-3515 or go online to www.youngliving.com to place your order. You must have an Enroller and/or Sponsor number. This is the person who shared this book with you or who first introduced you to Young Living Essential oils.

   - For online order, first select your country, and then click Become a Member.

   - You may choose to become a retail customer and pay full price or become a wholesale member and save 24%. There are no monthly obligations for purchasing or recruiting, and no yearly membership renewal as a wholesale member; this can simply be a wholesale account for your personal purchases. If you desire, you may initiate a home-based business, but it is never required.

   - If you select the retail customer option, make your purchases, and when the products arrive, enjoy!

   - If you elect wholesale member, select your kit. (The author suggests the Premium Starter Kit with Essential Oils as it contains many of the oils mentioned in this book.) Then, select Continue Shopping to order other products or complete the ordering process.

4. Be sure to write down your new Young Living member number, personal identification number, and password for future orders. Then, wait for your products to arrive and enjoy!

The CBD Isolate can be purchased through Nature's Ultra. This brand of CBD is infused with Young Living's therapeutic grade essential oils. This can be purchased at www.naturesultra.com You can purchase the CBD oil infused with Young Living Essential Oils, or you can get the unflavored and add your own essential oil combination to meet your needs.

BIBLIOGRAPHY

Stewart, David. *The Chemistry of Essential Oils Made Simple: God's Love Manifest in Molecules*. Missouri, CARE, 2010

Essential Oils Desk Reference, 5th Edition. Lifescience Publishing, 2011

1. Battista, Natalia et al. "Altered expression of type-1 and type-2 cannabinoid receptors in celiac disease" PloS one vol. 8,4 e62078. 19 Apr. 2013, doi:10.1371/journal.pone.0062078
2. Battista, Natalia et al. "Abnormal anandamide metabolism in celiac disease" The Journal of Nutritional Biochemistry Volume 23, Issue 10, October 2012, Pages 1245-1248
3. Pertwee, Roger G. "Cannabinoid pharmacology: the first 66 years" *British journal of pharmacology* vol. 147 Suppl 1,Suppl 1 (2006): S163-71.
4. Kreitzer, Anatol C. "Neurotransmitters: Emerging Role of Endocannabinoids" Current Biology Volume 15, Issue 14, 26 July 2005, Pages R549-R551
5. Atakan, Zerrin. "Cannabis, a complex plant: different compounds and different effects on individuals" *Therapeutic advances in psychopharmacology* vol. 2,6 (2012): 241-54.
6. Atakan, Zerrin. "Cannabis, a complex plant: different compounds and different effects on individuals" *Therapeutic advances in psychopharmacology* vol. 2,6 (2012): 241-54.
7. "Hemp & Flax in Colonial America." *Terms of Estrangement: Who Were the Sons of Liberty?: The Colonial Williamsburg Official History & Citizenship Site*, www.history.org/foundation/journal/winter15/hemp.cfm.
8. "George Washington Grew Hemp." *George Washington's Mount Vernon*, www.mountvernon.org/george-washington/facts/george-washington-grew-hemp.
9. "The Controlled Substances Act." DEA, www.dea.gov/controlled-substances-act.
10. "Agriculture Improvement Act of 2018 (2018 - H.R. 2)." *GovTrack.us*, www.govtrack.us/congress/bills/115/hr2.
11. "CBD History: The Past, Present And Future of Using CBD." *Ministry of Hemp*, Ministry of Hemp, 12 July 2018, ministryofhemp.com/blog/cbd-history

12. Zuardi, Antonio Waldo. (2006). History of cannabis as a medicine: a review. *Brazilian Journal of Psychiatry*, 28(2), 153-157. https://dx.doi.org/10.1590/S1516-44462006000200015
13. Goldstein, Bonni. "Cannabis in the Treatment of Pediatric Epilepsy." *TheRoc*, O'Shaughnessy's, 2015, www.theroc.us/researchlibrary/Cannabis%20in%20the%20Treatment%20of%20Pediatric%20Epilepsy.pdf.
14. "Extraction Methods: How Are CBD Oils & CBD Isolate Made?" *IntelliCBD*, 28 Sept. 2018, intellicbd.com/articles/how-is-cbd-made-extracted/.
15. "Everything You Need to Know About CBD Isolate." CBD School, Publisher Name *CBD School* Publisher Logo, 13 Feb. 2019, www.cbdschool.com/how-to-use-cbd-isolate/.
16. "Everything You Need to Know About CBD Isolate." *CBD School*, Publisher Name CBD School Publisher Logo, 13 Feb. 2019, www.cbdschool.com/how-to-use-cbd-isolate/.
17. Russo, Ethan B. "Taming THC: potential cannabis synergy and phytocannabinoid-terpenoid entourage effects" *British journal of pharmacology* vol. 163,7 (2011): 1344-64.
18. Luigi, Romano L. "Cannabis Oil: Chemical Evaluation of an Upcoming Cannabis-Based Medicine." *Cannabis-Med*, 2013, www.cannabis-med.org/data/pdf/en_2013_01_1.pdf.
19. Iffland, Kerstin and Franjo Grotenhermen. "An Update on Safety and Side Effects of Cannabidiol: A Review of Clinical Data and Relevant Animal Studies" *Cannabis and cannabinoid research* vol. 2,1 139-154. 1 Jun. 2017, doi:10.1089/can.2016.0034
20. Iffland, Kerstin and Franjo Grotenhermen. "An Update on Safety and Side Effects of Cannabidiol: A Review of Clinical Data and Relevant Animal Studies" *Cannabis and cannabinoid research* vol. 2,1 139-154. 1 Jun. 2017, doi:10.1089/can.2016.0034
21. Atakan, Zerrin. "Cannabis, a complex plant: different compounds and different effects on individuals" *Therapeutic advances in psychopharmacology* vol. 2,6 (2012): 241-54.
22. Costa, B et al. "The non-psychoactive cannabis constituent cannabidiol is an orally effective therapeutic agent in rat chronic inflammatory and neuropathic pain." Eur J Pharmacol.2007 Feb 5;556(1-3):75-83. Epub 2006 Nov 10.
23. Genaro, Karina et al. "Cannabidiol Is a Potential Therapeutic for the Affective-Motivational Dimension of Incision Pain in Rats" *Frontiers in pharmacology* vol. 8 391. 21 Jun. 2017, doi:10.3389/fphar.2017.00391

24. Federation of American Societies for Experimental Biology. "'Marijuana receptor' might hold the key to new fertility treatments for men: Research suggests that cannabis exposure may affect DNA-bound proteins, sperm chromatin and have an impact on fertility, embryo development and offspring health." ScienceDaily. ScienceDaily, 8 April 2016. <www.sciencedaily.com/releases/2016/04/160408163828.htm>.

25. "The Clinical Significance of Endocannabinoids in Endometriosis Pain Management." *Mary Ann Liebert, Inc. | Home*, www.liebertpub.com/doi/full/10.1089/can.2016.0035.

26. Bergamaschi, Mateus M et al. "Cannabidiol reduces the anxiety induced by simulated public speaking in treatment-naïve social phobia patients" *Neuropsychopharmacology : official publication of the American College of Neuropsychopharmacology* vol. 36,6 (2011): 1219-26.

27. Iseger, Tabitha A. et al. "A systematic review of the antipsychotic properties of cannabidiol in humans" Schizophrenia Research Volume 162, Issues 1–3, March 2015, Pages 153-161

28. Zanelati, T V et al. "Antidepressant-like effects of cannabidiol in mice: possible involvement of 5-HT1A receptors" *British journal of pharmacology* vol. 159,1 (2009): 122-8.

29. Long, Leonora E et al. "Distinct neurobehavioural effects of cannabidiol in transmembrane domain neuregulin 1 mutant mice" PloS one vol. 7,4 (2012): e34129.

30. Prud'homme, Mélissa et al. "Cannabidiol as an Intervention for Addictive Behaviors: A Systematic Review of the Evidence" *Substance abuse: research and treatment* vol. 9 33-8. 21 May. 2015, doi:10.4137/SART.S25081

31. Weiss, Lola et al. "Cannabidiol arrests onset of autoimmune diabetes in NOD mice" *Neuropharmacology* vol. 54,1 (2007): 244-9.

32. L. Weiss, M. Zeira, S. Reich, M. Har-Noy, R. Mechoulam, S. Slavin & R. Gallily (2006) Cannabidiol lowers incidence of diabetes in non-obese diabetic mice, Autoimmunity, 39:2, 143-151, DOI: 10.1080/08916930500356674

33. Jadoon, Khalid A et al. "A single dose of cannabidiol reduces blood pressure in healthy volunteers in a randomized crossover study" *JCI insight*, vol. 2,12 e93760. 15 Jun. 2017, doi:10.1172/jci.insight.93760

34. Rajesh, Mohanraj et al. "Cannabidiol attenuates cardiac dysfunction, oxidative stress, fibrosis, and inflammatory and cell death signaling pathways in diabetic cardiomyopathy" *Journal of the American College of Cardiology* vol. 56,25 (2010): 2115-25.

35. Shannon, Scott and Janet Opila-Lehman. "Effectiveness of Cannabidiol Oil for Pediatric Anxiety and Insomnia as Part of Posttraumatic Stress Disorder: A Case Report" *Permanente journal* vol. 20,4 (2016): 108-111.

36. Olah, A et al. "Cannabidiol exerts sebostatic and antiinflammatory effects on human sebocytes." *J Clin Invest*. 2014 Sep;124(9):3713-24. doi: 10.1172/JCI64628. Epub 2014 Jul 25.

37. Olah, A. et al. "Differential effectiveness of selected non-psychotropic phytocannabinoids on human sebocyte functions implicates their introduction in dry/seborrhoeic skin and acne treatment" *Experimental Dermatology* Volume 25, Issue 9, September 2016 Pages 701-707

38. Devinksy, Orrin et al. "Trial of Cannabidiol for Drug-Resistant Seizures in the Dravet Syndrome" May 25, 2017 N Engl J Med 2017; 376:2011-2020 DOI: 10.1056/NEJMoa1611618

39. Chagas, MH et al. "Effects of cannabidiol in the treatment of patients with Parkinson's disease: an exploratory double-blind trial."*J Psychopharmacol*. 2014 Nov;28(11):1088-98. doi: 10.1177/0269881114550355. Epub 2014 Sep 18.

40. Chagas, MH et al. "Cannabidiol can improve complex sleep-related behaviours associated with rapid eye movement sleep behaviour disorder in Parkinson's disease patients: a case series." *J Clin Pharm Ther*. 2014 Oct;39(5):564-6. doi: 10.1111/jcpt.12179. Epub 2014 May 21.

41. Chang, D et al. "Long-term cannabidiol treatment prevents the development of social recognition memory deficits in Alzheimer's disease transgenic mice." *J Alzheimers Dis*. 2014;42(4):1383-96. doi: 10.3233/JAD-140921.

42. Cheng, D et al. "Long-term cannabidiol treatment prevents the development of social recognition memory deficits in Alzheimer's disease transgenic mice." *J Alzheimers Dis*. 2014;42(4):1383-96. doi: 10.3233/JAD-140921.

43. Bloechl-Daum, B et al. "Delayed nausea and vomiting continue to reduce patients' quality of life after highly and moderately emetogenic chemotherapy despite antiemetic treatment." *J Clin Oncol*. 2006 Sep 20;24(27):4472-8.I. Fraguas-Sánchez, A. Fernández-Carballido & A. I. Torres-Suárez (2016) Phyto-, endo- and synthetic cannabinoids: promising chemotherapeutic agents in the treatment of breast and prostate carcinomas, Expert Opinion on Investigational Drugs, 25:11, 1311-1323, DOI: 10.1080/13543784.2016.1236913

44. A. I. Fraguas-Sánchez, A. Fernández-Carballido & A. I. Torres-Suárez (2016) Phyto-, endo- and synthetic cannabinoids: promising chemotherapeutic agents in the treatment of breast and prostate carcinomas, Expert Opinion on Investigational Drugs, 25:11, 1311-1323, DOI: 10.1080/13543784.2016.1236913

45. McAllister, SD et al. "Cannabidiol as a novel inhibitor of Id-1 gene expression in aggressive breast cancer cells." *Mol Cancer Ther*. 2007 Nov;6(11):2921-7.

46. Cannabidiol Induces Programmed Cell Death in Breast Cancer Cells by Coordinating the Cross-talk between Apoptosis and Autophagy Ashutosh Shrivastava, Paula M. Kuzontkoski, Jerome E. Groopman and Anil Prasad *Mol Cancer Ther* July 1 2011 (10) (7) 1161-1172; DOI 10.1158/1535-7163.MCT-10-1100

47. Uberto Pagotto, Giovanni Marsicano, Daniela Cota, Beat Lutz, Renato Pasquali; The Emerging Role of the Endocannabinoid System in Endocrine Regulation and Energy Balance, *Endocrine Reviews*, Volume 27, Issue 1, 1 February 2006, Pages 73–100, https://doi.org/10.1210/er.2005-0009

48. Anand, Praveen et al. "Targeting CB2 receptors and the endocannabinoid system for the treatment of pain" *Brain research reviews* vol. 60,1 (2008): 255-66.

49. Ruehle, S., et al. "The Endocannabinoid System in Anxiety, Fear Memory and Habituation." *Journal of Psychopharmacology*, vol. 26, no. 1, Jan. 2012, pp. 23–39, doi:10.1177/0269881111408958.

50. Ibrahim, Mohab M et al. "Activation of CB2 cannabinoid receptors by AM1241 inhibits experimental neuropathic pain: pain inhibition by receptors not present in the CNS" *Proceedings of the National Academy of Sciences of the United States of America* vol. 100,18 (2003): 10529-33.

51. Silvestri, C et al. "The Endocannabinoid System in Energy Homeostasis and the Etiopathology of Metabolic Disorders"Cell Metabolism|VOLUME 17, ISSUE 4, P475-490, APRIL 02, 2013 DOI:https://doi.org/10.1016/j.cmet.2013.03.001

52. Pacher, Pál et al. "Modulation of the endocannabinoid system in cardiovascular disease: therapeutic potential and limitations" *Hypertension (Dallas, Tex.: 1979)* vol. 52,4 (2008): 601-7.